i

UNDERSTANDING AND USING THE STEREOMICROSCOPE

Lewis Woolnough

2nd Edition, 2018

Previously published in 2010 as
ISBN 978-0-9564591-0-7
by the Quekett Microscopical Club

This edition published by Milton Contact Ltd.

A CIP catalogue record for this book is available
from the British Library.

ISBN 978-1-911526-23-0

Milton Contact Ltd
3 Hall End, Milton, Cambridge CB24 6AQ

Preface to Revised Edition

Behind this book was the perceived need for an uncomplicated but thorough introduction to stereomicroscopy. It seeks to provide an understanding of basic theory and engagement with the issues involved in getting the best results when using a stereomicroscope. A considerable amount of relevant material is scattered throughout the literature and it is hoped that organising the most significant of this in one publication will be helpful.

The stereomicroscope is sometimes considered to be less useful than a conventional "compound" instrument. This is unfortunate because it is certainly not a kind of poor relation or cheap alternative. Each type of microscope has its limitations but excels in certain circumstances; they complement each other.

Whilst it is comparatively easy to get an image with a stereomicroscope, a basic understanding of theoretical and practical issues is needed if the best results and a range of stunning effects are to be obtained.

This book will assist both established and aspiring stereomicroscopists to get excellent results with their instruments and have much satisfaction and enjoyment in the process.

It is clear that many readers of the first edition of this book would have appreciated more guidance about taking photographs through the stereomicroscope and an extra chapter has been included here with a view to meeting that need. The list of references has been expanded and there are some new illustrations but it has not been deemed necessary to make any fundamental changes to the book's content.

Lewis Woolnough,
Little Whelnetham,
Bury St. Edmunds.

March, 2018

Contents

Introduction

Stereomicroscopes are designed to give erect and naturally aligned images with a sense of depth.

Magnifications are typically in the range x10 to x100.

Working distances are large.

These instruments are, therefore, invaluable for preliminary examination, manipulation and dissection of specimens.

Compact Disk

Quickstart

This section is primarily for those who have no experience of using the stereomicroscope or who wish to try out an instrument.

Use the manufacturer's instructions to assemble a new microscope. Second-hand instruments are usually ready for use but the illustrations in this book will give guidance, if necessary.

Follow the four-step process on the next page.

What do you observe?

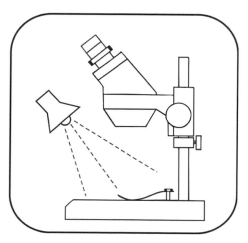

1. Place the instrument where there is plenty of light from a window or illuminate with a desk lamp, as indicated.

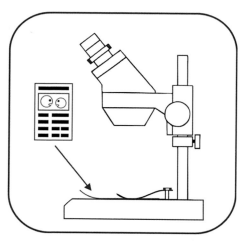

2. Position a newspaper photograph onto the microscope *stage*, as shown.

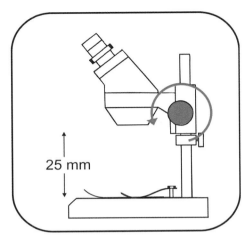

25 mm

3. If there are *lenses* with different magnifications, move the one with the lowest magnification into position.

<u>Looking from the side</u>, turn the arrowed knob to lower the lenses until they are about 25 mm from the newspaper, as shown in the diagram, or as low as possible.

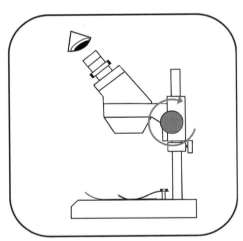

4. Look through the *eyepieces*. Adjust the distance between them, much as one would do with a pair of field binoculars, for comfortable viewing.

Then, turn the *focussing knob* again, this time to raise the lenses, until you can see the magnified photograph clearly and in focus.*

*It might be possible to focus one, or both, of the eyepiece lenses independently. You can make adjustments, if necessary, by looking with one eye at a time and twisting any rings available on the respective lens tubes. You should get a very clear image with both eyes after making small movements of the main focussing knob.

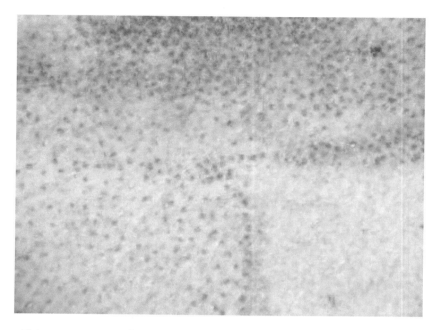

This newspaper photograph is made up of numerous dots and the density of the dots is important.

With a black and white photograph, a white area has no dots; shades of grey are created by various densities of dots; in black areas, the dots are so tightly packed that the paper surface is covered with black ink.

In a colour photograph, there are dots of four colours: magenta, cyan, yellow and black. The different combinations of these and their distribution give the colour effects on the surface of the print.

Please move on for more about how stereomicroscopes work and how to get the best results from them. You will also find suggestions about interesting things to examine and places to look for further information.

Quickstart

Examples of things to look at

Butterfly Scales, Moss Capsules,
Moss Leaves, Aphid,
Sand, Foraminiferans.

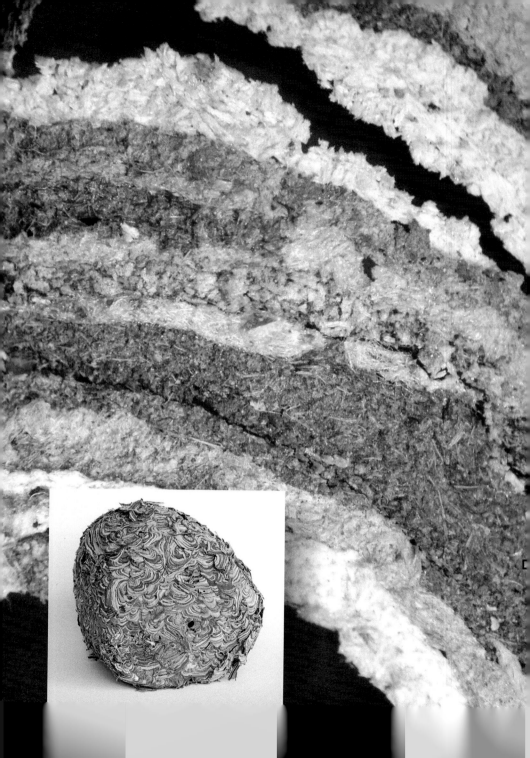

General Principles

A complex series of events takes place to make it possible for us to see an object.

Some of the light given off by the object (or reflected from it) passes into the eye and through its *lens* to create an image on the internal surface at the back of the eye, the *retina*. The stimulation of nerve endings at this surface causes electrical impulses to pass along the *optic nerve* to the *brain*, which processes them to produce the sensation that is called "seeing the object".

The illustrations on the following pages illustrate this process and also deal with the importance of perception and interpretation."

The Process of Seeing an Object

As the diagram suggests, the brain creates a sensation of the object being correctly represented in terms of its top / bottom and left / right orientation. However, it is also capable of considerable "accommodation" with regard to a number of features; this means, for example, that all parts of an object might appear in focus whereas a photograph would show only some parts as being really sharp.

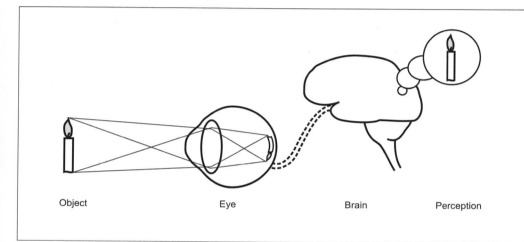

| Object | Eye | Brain | Perception |

Not to scale.

Experience contributes to the brain's ability to produce coherent and reliable interpretation of scenes that we observe. It is clearly possible for us to develop a sense of depth in a scene, even using one eye, by drawing upon our knowledge of the relative sizes of objects, the planes in which they are apparently positioned, etc.

Try this.

What do you see in this picture?

Move on to the next page when you have decided.

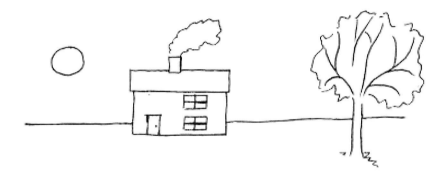

Now look at the picture below.

Do you still see the same objects that were in the original picture?

Extra information has demanded a re-interpretation of the scene! Where before you might have thought you were seeing the sun shining on a house and tree, you now see a boy throwing a ball at a different time of day.

So, we can be deceived. Sometimes, what we think we see is not what is actually there!

The same thing applies when we interpret the images we see when using a microscope and some caution is needed. An example of an artefact that can cause confusion is the simple air bubble trapped in a liquid or resin mountant; it appears to be an interesting black circular structure!

Air Bubbles Trapped in Tiny Moss Leaves in Water.

General Principles

Let us now consider the observation of a simple object by two eyes. This is STEREOSCOPIC VISION.

The distance between the two eyes of an adult person is about 6.5 cm. Each eye will give its own view of an observed object, the difference being more marked with closer, than with more-distant, objects.

Try this.

Hold a book vertically in front of your face so that you are looking at the spine which is about 10 cm from your nose.

Keeping the book and your head still, close one eye. What do you see?

Now view only with the eye that was closed. What do you see now?

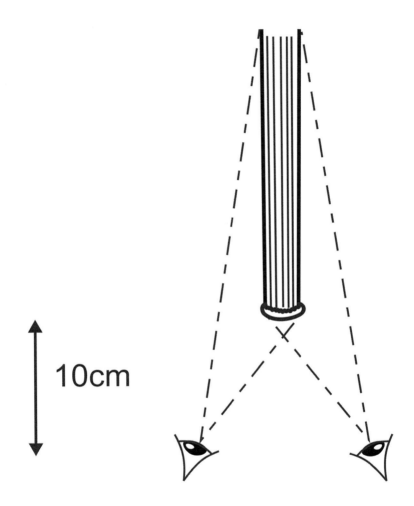

Top view:
A slightly different image of the book is seen by each eye.

Move on to the next page for an explanation.

General Principles

It is clear that each eye sees something different. The brain, receiving signals from both eyes, combines the information to produce a sensation of seeing an object that appears to have depth - it is three-dimensional.

Try this.

(For some people, it is easy to achieve the desired result; others find that it takes some time and effort. A few may conclude that it cannot be done but difficulty with this does not mean that the use of the stereomicroscope will be a problem).

Look at these images opposite at a distance of about 20 – 30 cm. Try to look into the far distance beyond the space between the images. Copies of the images should appear to move together to form a third image which is visible between the drawn ones; your attention should gradually become concentrated on the new image which assumes dominance as the drawn images become less distinct.

Does the new image have a three-dimensional quality?

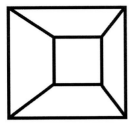 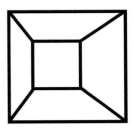

There should appear to be a truncated pyramid coming out of the paper <u>towards you</u>.*

This is a *stereoscopic* effect which may be considered to be a true stereoscopic (*orthostereoscopic*) image.

*What is described here is what is known as *'Parallel Viewing'*. Some people will automatically cross their eyes to get the 3D effect. This is termed *'Cross-eyed Viewing'*.

Repeat what you just did.
(This time, the diagrams have been transposed.)

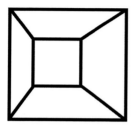 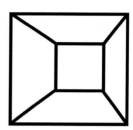

What do you see this time?

The truncated pyramid seems to be going down below the plane of the paper, <u>away from you</u>.*

This is a *pseudostereoscopic* effect.

* For some people, the 3D effect may be reversed or interchangeable.

General Principles

Stereoscopic vision is one of the delights of using the Stereomicroscope! Try the viewing technique that you have just been using with the two images below which, at first glance, appear to be identical.

Do you get a better view of your subject or a poorer one?

Genital Capsule of a Male Bumble bee, <u>Bombus pascuorum.</u>

Insect Preserved in Amber.

Potassium Permanganate Crystals.

General Principles

Anthers (Stained) of Meadow Crane's-bill, <u>Geranium pratense</u>.

Stereoscopes and Their Use

The *stereoscope* is a device used for viewing *stereo-pairs* of photographs. Essentially, there is a pair of lenses, one for each eye, through which one views the pair of photographs that are placed side by side; the two images need to be mounted so that any two equivalent points are the same distance above a common base-line and the distance between such points should not exceed the *inter-pupillary distance* for the viewer, viz. the distance between the eyes.

Antique stereoscopes are available but there are also modern versions on the market and it is not difficult to construct an effective stereoscope that will enable everyone with normal eyesight to enjoy stereoscopic effects comfortably (See illustrations below and opposite).

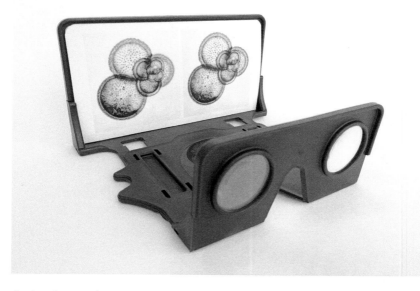

A simple, modern stereoscope that can be purchased new from The London Stereoscopic Company.

A home-made stereoscope. The lenses are from an inexpensive pair of binoculars. It is possible to adjust the distance of the photographs from the lenses and the distance between the lenses.

General Principles

Delphinium Flower

How Microscopes Work

A microscope is a device for creating an image which shows, to a greater or lesser extent, the details in the structure of an object. This image must be made large enough to be seen easily.

Types of Microscope

For practical purposes, there are three types of microscope:
1. The Simple Microscope
2. The Compound Microscope
3. The Stereomicroscope.

1. The Simple Microscope

This is really what we would recognise as a magnifying glass, a single lens that may be hand-held or mounted on a stand. The lens may have more than one glass element; this is to reduce certain distortions and undesirable colour effects (aberrations) that would arise with a single-element lens.

2. The Conventional Compound Microscope

The picture opposite shows an instrument that has two lenses, one at the bottom and one at the top of a tube, working together. An *objective lens*, close to the object, creates a magnified version of it (or part of it). A second lens, the *eyepiece*, is the one through which the observer looks; this enlarges the first image a bit more and modifies the light ready for the eye to receive it.

There is one eyepiece in this *monocular* instrument. One can view with both eyes at once with the *binocular* version; there are two eyepieces but the important point here is that each eye will see an identical image of the object.

This type of microscope, whether a monocular or binocular version, will not create a stereoscopic image.

The following pages give more detail about the way "Compound" microscopes work. This section deals with fundamental concepts of image formation but avoids the use of complicated mathematics. More detailed treatment of this subject can be found in many textbooks, some of which are listed in "References and Further Reading."

You can omit this section if you wish and go straight on to page 32.

How Microscopes Work

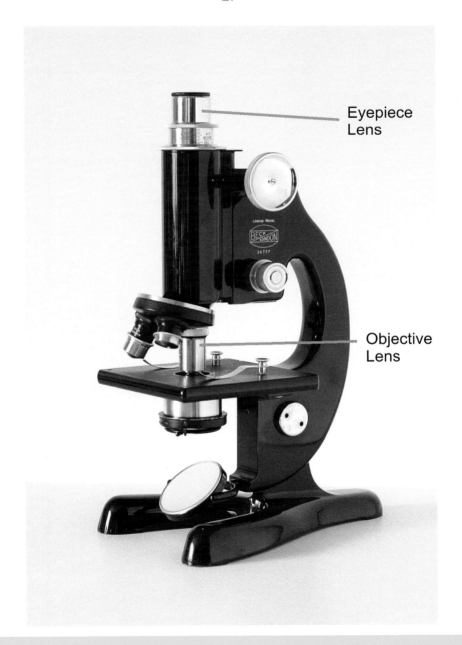

Eyepiece
Lens

Objective
Lens

How Microscopes Work

2. The Conventional Compound Microscope, cont.

A *monocular compound microscope* is for viewing with one eye. It has two essential elements, the *objective lens* complex and the *eyepiece lens* complex. These are at the bottom and top, respectively, of a tube.

The objective lens forms a *primary image* which is a magnified version of the object (or part of it). The eyepiece provides some further enlargement of this image and modifies the light passing through for reception by the eye of the observer.

The diagram opposite shows the essentials of the process of image formation by the objective and eyepiece lenses and an observing eye. Notice the re-orientation of the image at the different planes in the system. The orientation of the "observed" image is different from that of the object; movement of the object in any direction will be observed through the instrument as a movement in the opposite direction.

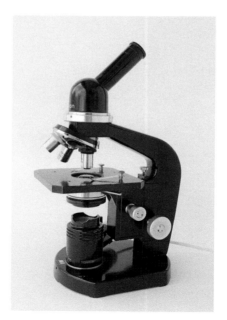

How Microscopes Work

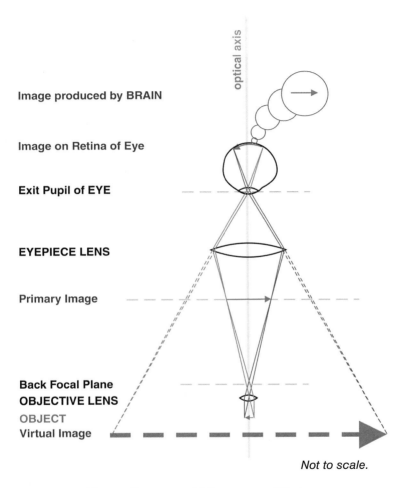

Not to scale.

How a Compound Microscope Works.
The blue and green lines show the paths of sample light rays from the object at the bottom of the figure to the eye at the top.

How Microscopes Work

2. The Conventional Compound Microscope, cont.

The *resolution* or amount of detail recorded in the primary image is determined by the action of the objective lens. A measure of this is the *minimum resolved distance*; this is the smallest distance between two points at which they can be seen as discrete, rather than appearing to be one single point.

Factors determining this are:

1. *Wavelength* of the light (λ). This can be taken as 0.55 µm.*

2. The *Numerical Aperture* (N.A.). This is a measure of the capacity of the objective lens to collect light. In practice, values range from less than 0.1 to about 1.3.

Minimum resolved distance (M.R.D.) is given by this formula attributed to Ernst Abbe (1840 – 1905) :

$$M.R.D. = 0.5 \times \lambda \div N.A.$$

Try this   The answers are given at the bottom of page 32.

Can you work out the M.R.D. when the N.A. has the following values?
a). 0.1 and
b). 1.3

*Note. 1 µm = 1 / 1,000,000 m or 1 / 1,000 mm.

How Microscopes Work

2. The Conventional Compound Microscope, cont.

It can be seen that the M.R.D. decreases (i.e. the detail resolved increases) as the N.A. increases.

A more powerful lens will tend to have a higher N.A. and will be much closer to the object when it is in focus. We say that it has a shorter *working distance*.

'Low Power'	**'High Power'**
Small NA	Large NA
Low Magnification	High Magnification
Large Working Distance	Small Working Distance
Greater Depth of Field	Smaller Depth of Field

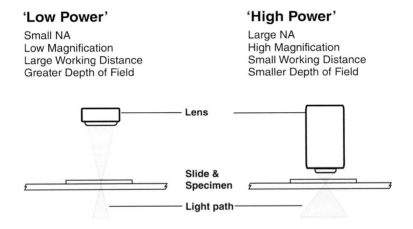

Illustration of low power and high power lens.

In the *binocular* version of this instrument, there are two eyepieces but each eye will see an identical image of the object. A stereoscopic effect is not created.

.

How Microscopes Work

3. The Stereomicroscope

Again, we have objectives and eyepieces working together but now there are always effectively two of each so that the instrument is essentially like two monocular compound microscopes, side by side; one serves the left eye, the other the right.

The key difference here is that there are optical arrangements so that <u>each eye sees a slightly different view (of the object)</u>. This is equivalent to straightforward observation of an object with both eyes and we have seen that an impression of its three-dimensional quality is created.

Stereomicroscopes give a considerable *depth of vision* (the extent of the thickness of a specimen that can be seen clearly) and this contributes significantly to the creation of a three-dimensional effect in what is seen.

(Answers to challenges on page 30: a). 2.75 µm b). 0.21 µm)

How Microscopes Work

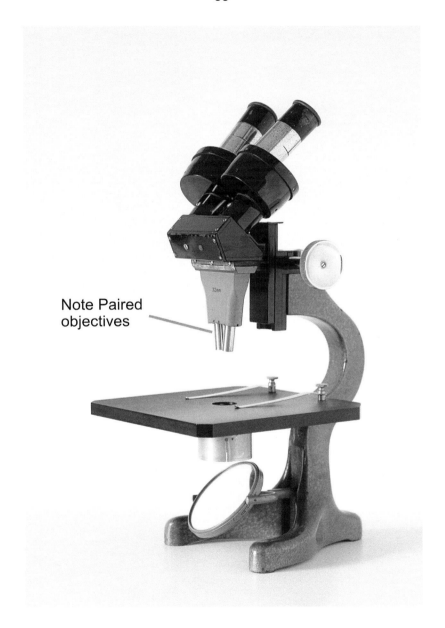

Note Paired
objectives

Vision With Microscopes

1. The Conventional Compound Microscope

A monocular microscope or a binocular instrument in which both eyes see the same image has a limited capacity to give a sense of an object's depth.

It has been noted that the depth of vision contributes to the creation of a three-dimensional effect; it is determined by the *focal depth* of the optical system and the accommodation of the eye (which will vary from person to person). As the *power* (numerical aperture and, generally, magnification) of the objective lens is increased, the depth of vision decreases and the contribution made by eye accommodation becomes increasingly less significant.

However, the important point here is that, with these microscopes, all of the image that appears sharp seems to lie in the same, single plane. There is a very limited sense that the specimen has depth.

How is the situation different with stereomicroscopes?

How Microscopes Work

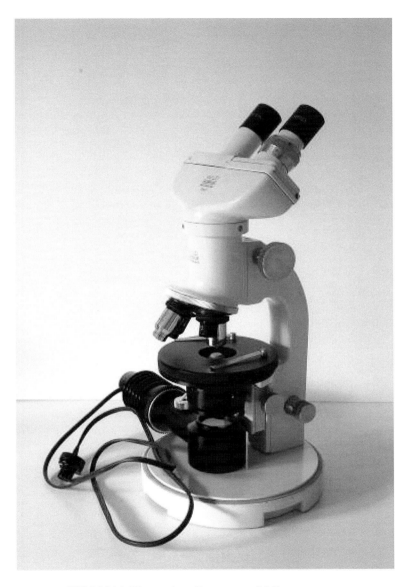

Wild M11 Binocular, Compound Microscope -
This DOES NOT give a stereoscopic effect.

How Microscopes Work

Vision With Microscopes, cont.

2. The Stereomicroscope

We have seen that, in straightforward observation of a three-dimensional object with both of our eyes, we get a sense of its depth. A binocular microscope system of optics in which <u>a different image of the specimen is provided for each eye</u> can create the same sort of effect very convincingly.

A pair of *Porro prisms*, positioned in each light path, ensures that the image observed has the same alignment as the object (see diagrams on pages 41 & 43). This is a key feature that makes the stereomicroscope the instrument of choice for dissection and manipulative work.

Stereoscopic microscopes operate with objective lenses of comparatively low power (low numerical apertures with correspondingly low magnifications). This means that there is a considerable depth of vision which contributes significantly towards the creation of a three-dimensional effect in what is observed.

If you worked through pages 28-31, it may be of interest to note that the N.A. provided by stereomicroscope objectives is typically less than 0.1. Applying calculations as before shows that the minimum resolved distance is about 3.00 µm or more. It is fair to say that the stereomicroscopist's main concern is not the revelation of extremely fine detail in an object!

How Microscopes Work

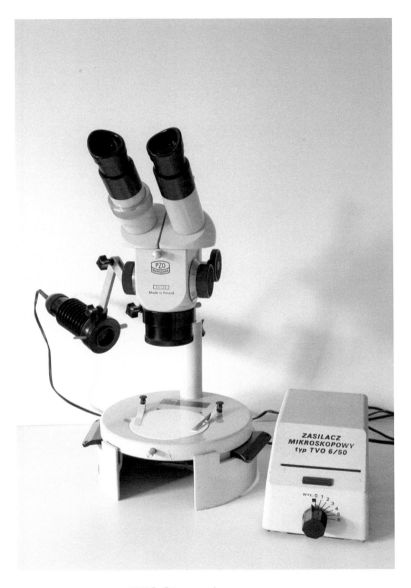

PZO Stereomicroscope –
This DOES give a stereoscopic effect.

How Microscopes Work

Stereomicroscope Systems: The Instruments

Most instruments now available and those currently manufactured use one or other of the following systems, upon which we shall now concentrate.

1. The Greenough System

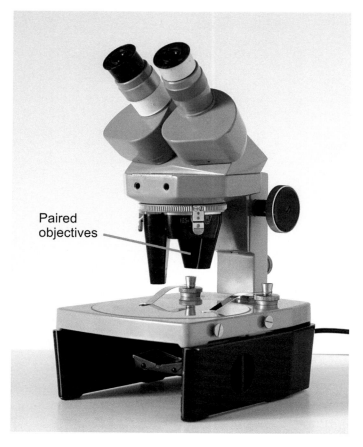

Paired objectives

Vickers Stereomicroscope.

How Microscopes Work

2. The Common Main Objective System

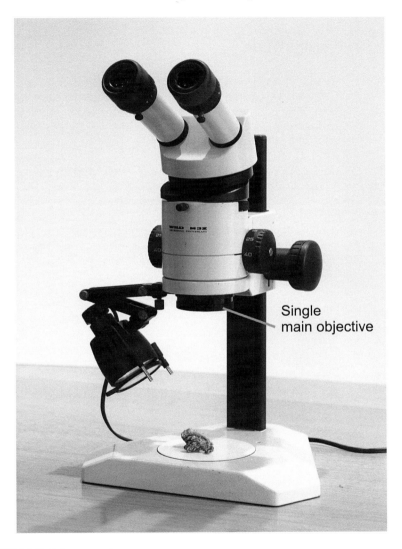

Single
main objective

Wild M3Z Stereomicroscope.

How Microscopes Work

Stereomicroscope Systems: Light paths

The Greenough Type

Here we have, effectively, two separate imaging systems, one for each eye. The *optical axes* of these are inclined to each other at an angle of about 16°, giving the two different viewpoints required to produce stereoscopic vision. The paired objective lenses are an obvious feature of this type of instrument.

As the diagram opposite shows, the axes of these optical paths are not perpendicular to the image plane; this is, theoretically, a problem for creating an image that is in focus across the whole *field of view*. However, the brain is capable of selecting only the sharp elements of each image for combination into the one that is "observed".

Zeiss started manufacturing microscopes to this design in 1897.

How Microscopes Work

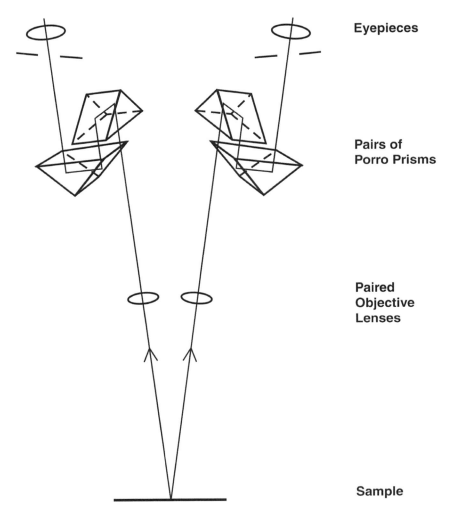

Eyepieces

**Pairs of
Porro Prisms**

**Paired
Objective
Lenses**

Sample

Light paths in a Greenough Stereomicroscope.

From Bradbury and Bracegirdle, with permission.

How Microscopes Work

Stereomicroscope Systems: Light paths Cont.

The Common Main Objective (C.M.O.) Type

These instruments provide the two optical paths, one for each eye and giving the two viewpoints required, by using <u>one</u> front (objective) lens.

With this approach, the optical paths emerging from the system are parallel to one another and both perpendicular to the image plane.

Again, Zeiss led the way and started production of this type of instrument in 1946.

How Microscopes Work

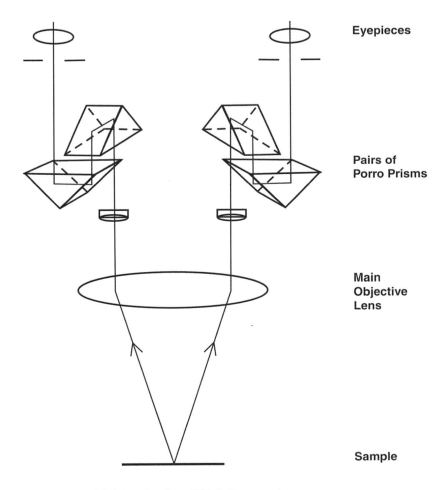

Eyepieces

Pairs of
Porro Prisms

Main
Objective
Lens

Sample

Light paths in a CMO Stereomicroscope.

From Bradbury and Bracegirdle, with permission.

How Microscopes Work

Stereomicroscope Systems: Positive Features Compared

Greenough

- The central part of the objective lens is used for image formation.

- Higher numerical apertures can be achieved.

- Lens aberrations can be more easily corrected.

- The field of view is relatively large.

C.M.O.

- One can expect true focus across the entire field of view.

- With the ray paths parallel through most of the instrument, it is easier to incorporate features such as supplementary magnification changers (including *zoom optics*), polarisation accessories, etc.

- Provision for photomicrography can more easily be made.

How Microscopes Work

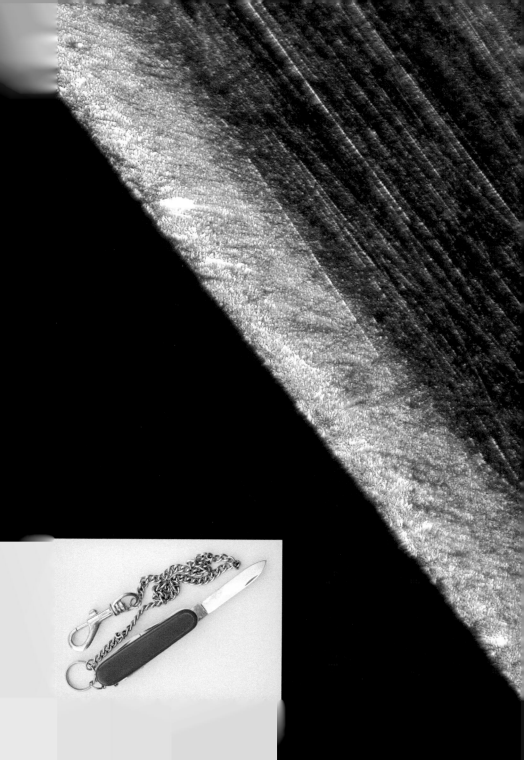

The Stereomicroscope's Features

A stereomicroscopy unit is built around a box-like structure that contains the main optical components. This is often referred to as the *head*.

The head will be supported on a *stand* that extends up from a solid supporting *base*.

There will be a means of moving the head up and down along its vertical axis, usually with a *focussing knob*, so that the position of the objective(s) relative to the specimen can be altered.

The objective lens, or lenses, will be at the underside of the head. Each eyepiece lens will sit in a tube on the upper part of the head. The tubes are moveable so that the distance between the eyepieces (*inter-ocular distance*) can be adjusted for comfortable viewing, much as one would do with a pair of field binoculars.

The base may be a simple platform, often with *stage clips* to steady mounted material, or it may incorporate equipment (*mirror* & / or *lamp*) for the provision of "bottom lighting" through the specimen from below. Some instruments are designed for the specimen to be positioned on a *stage*; this is a horizontal metallic or glass platform between the base and the objective, as with a conventional microscope (see illustration on page 33).

Lighting from above may be available from equipment attached to the stand or head.

Example: The Head of a Bausch and Lomb Stereomicroscope

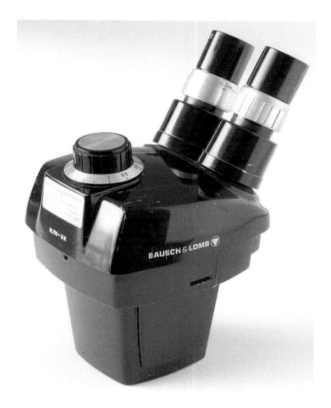

This is the head (or "Pod") of a Bausch and Lomb Stereo-Zoom.

The photographs on the following pages show it mounted in four of the different bases available for it.

The Stereomicroscope's Features

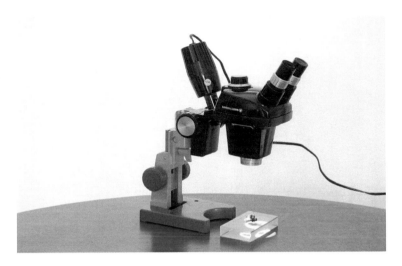

1. Simple base. Lamp for top lighting supported on stand.

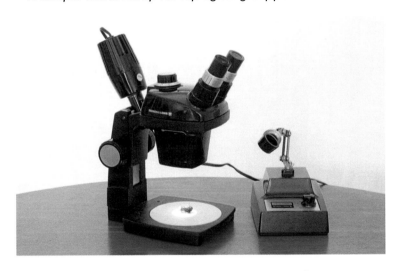

2. Base with white disc to support specimen.
(The reverse side of the disc is black).
Lamp for top lighting on stand; separate lamp stand available.

The Stereomicroscope's Features

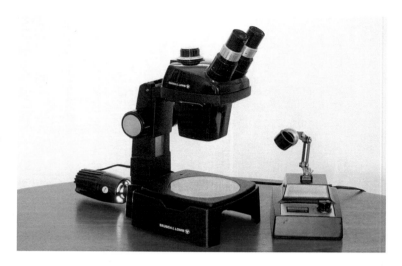

3. Stand with lamp inserted to give lighting from below. Adjustable mirror below 'frosted glass' disc which supports specimen.

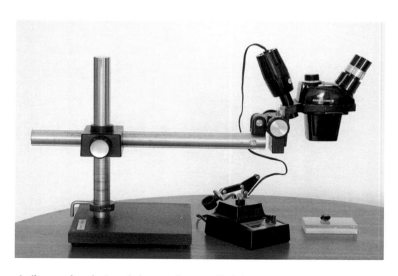

4. 'Long Arm' stand. Lamp for top lighting supported on stand.

The Stereomicroscope's Features

The Parts of a Basic Stereomicroscope

Try this.

Can you identify the parts indicated on the diagram below? It should be possible to find all of these features on any stereomicroscope.

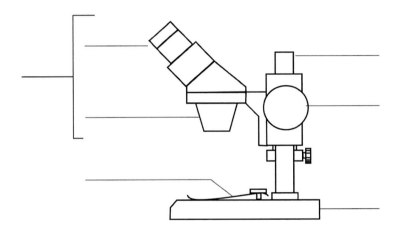

Turn to the next page for the answers.

Answers:
The Parts of a Basic Stereomicroscope

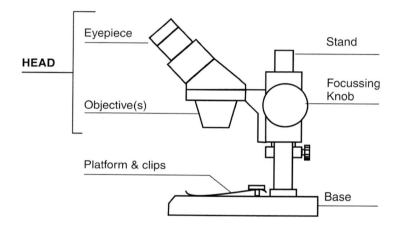

Eyepiece

Stand

HEAD

Focussing
Knob

Objective(s)

Platform & clips

Base

The Stereomicroscope's Features

Your eyepieces may be over the stand. Does the top part of the head (carrying the eyepieces) rotate on your instrument?

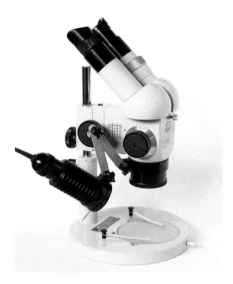

The Stereomicroscope's Features

Additional Features

Your instrument may well have a number of additional features. It is worth reading the manufacturer's instructions if these are available!

Try this.

Can you identify the following features on the microscope in this picture?

Attached lamp for top lighting.

Wheel in base for controlling the intensity of the light.

'Zoom' magnification changing knob.

Port for a tube to which a camera may be attached. With this tube, we have what is described as a *'trinocular head'*.

An eyepiece that can be focussed independently.

Which, if any, of these extra features does your microscope have?

Turn over to check your answers.

The Stereomicroscope's Features

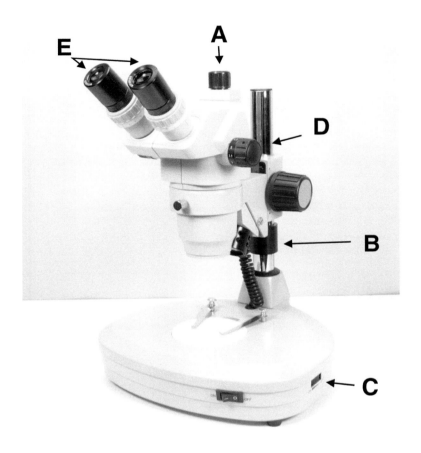

Brunel BMZ Zoom Stereomicroscope
(Stand ST2, with Illuminated Base).

From Brunel Microscopes, with permission.

The Stereomicroscope's Features

You will, no doubt, have noticed that this is a Common Main Objective Microscope? The parts labelled are as follows:

A. Port for a camera tube.

In most cases, one of the images created by the stereomicroscope is re-directed to the *phototube* on which the camera is attached; this is achieved by turning a knob or moving a slider. The sharp image one observes in normal viewing through the one eyepiece would then be captured by the camera as a non-stereomicroscopic image.
There are instruments available which enable the operator to photograph both images, in turn; these can be reproduced as stereo-pairs for use with a viewer, giving an apparently three-dimensional image. Such instruments are not common.

B. Lamp for top illumination.

C. Wheel in base for controlling light intensity.

D. "Zoom" magnification changer.

E. Eyepiece.

On this particular instrument, both eyepieces can be focussed independently. Having at least one eyepiece that can be adjusted independently, enables the operator to compensate for differences between the eyes and also to ensure that a sharp image of any *measuring graticule* inserted in the eyepiece is superimposed on the image of the object. More detail can be found on pages 80 & 81.

There may also be a *rotating stage*, *filter carriers*, etc. However, we shall now concentrate on the essentials of good technique that will help you to get good results from your stereomicroscope.

The Stereomicroscope's Features

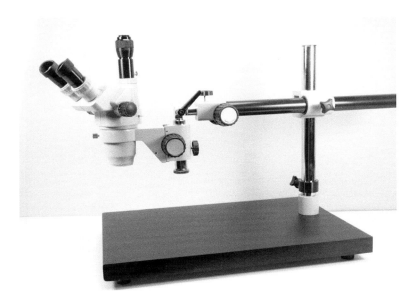

*Brunel BMZ Zoom Stereomicroscope
(Stand ST2, with Long-Arm Support).*

From Brunel Microscopes, with permission.

The Stereomicroscope's Features

Feather

Lighting

The illumination that can be provided will obviously be dependent upon the lighting equipment that one has or can obtain.

From the many options available, a minimum requirement is the provision of some form of *transmitted* ("from below") and *incident* ("top") lighting.

Transmitted light is used to reveal detail in transparent or translucent specimens which are usually comparatively thin.

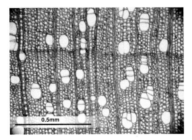

Silver Birch, Betula pendula.
T.S. Stem. Stained with safranin and light green.

Incident light will show up the general form and surface features of opaque specimens, some of which may be very thick.

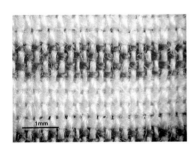

Woven threads of handkerchief.

NOTE:
The photomicrographs on this and page 61 were captured by a digital camera through **one** eyepiece of a Greenough type microscope, so the images are not stereoscopic.

Example of incident illumination, using a ring light attached to the stereomicroscope.

From GT Vision, with permission.

Experiment with Lighting

Sometimes it is helpful to use both transmitted light and incident light together. It is worth experimenting, especially with the relative intensity being provided by each source. There can be fewer complications if the same type of light (see next section) is provided from both directions.

These photomicrographs of the genital capsule of a bumblebee have all been taken in light from tungsten lamps. The same specimen has been used with different lighting arrangements.

When using incident lighting (lighting from above), it is worth experimenting; try illuminating the sample from different angles to achieve the best effect.

Lighting

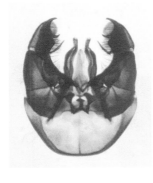

Transmitted light

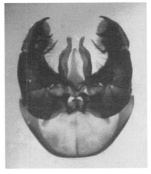

Incident light

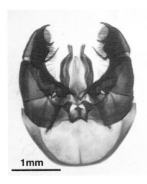

Incident AND transmitted light

Genital capsule of male bumblebee, <u>Bombus pascuorum</u>.

Lighting

Light Quality

Now, it will be helpful to turn our attention to LIGHT QUALITY

Our eyes can detect only some of the *electromagnetic radiation* that is all around us and this is described as the *visible spectrum.*

The 'quality of light' produced by a source is taken to mean the particular blend of wavelengths from the visible spectrum that are present. This will determine the colour of the light as it appears to us. It is not of great significance for straight observation of specimens; however, it is important for those taking photographs through a microscope to take this factor into account.

Try this.

Can you fill in the rainbow colours on the diagram of the visible spectrum on the facing page?

Turn to the next page for the correct sequence of colours.

Lighting

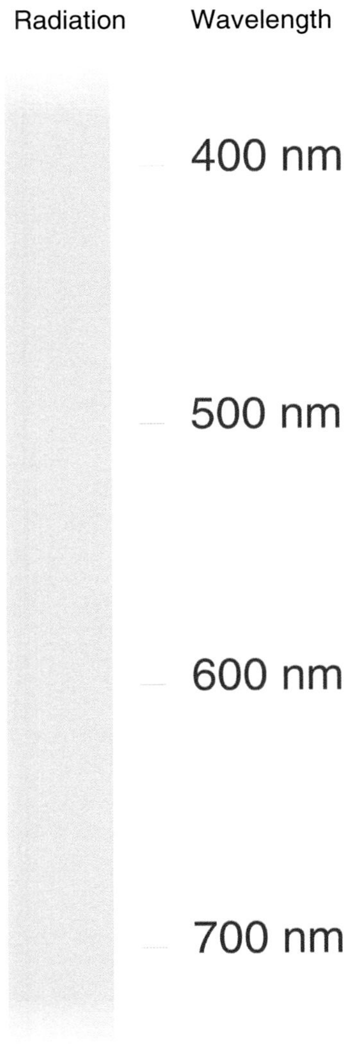

Radiation	Wavelength

400 nm

500 nm

600 nm

700 nm

1 nm = 1/1,000,000,000 m or 1/1,000,000 mm or 1/1000 μm

Lighting

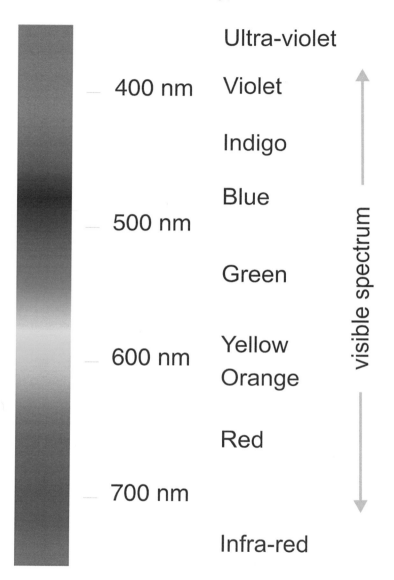

Ultra-violet

400 nm   Violet

Indigo

Blue

500 nm

Green

600 nm   Yellow
Orange

Red

700 nm

Infra-red

visible spectrum

Lighting

Common types of Artificial Illumination

Tungsten and Energy Saver Lamps

A significant proportion of the light produced by tungsten lamps is from the longer wavelengths section of the visible spectrum and appears to have a reddish tinge. Furthermore, with any increase in the voltage supplied (to increase the light intensity), the blend of radiation wavelengths produced will change, giving a bluer light. With age, the output of a tungsten lamp declines considerably. Energy saver lamps for domestic use give a more balanced "white" light.

Tungsten-Halogen Lamps

When operated at the appropriate voltage, these lamps give a bluer light than the tungsten type. A very bright light is produced and the continuous spectrum of wavelengths emitted changes very little throughout the lamp's life. The small filament is enclosed in a silica-glass envelope which will not soften at the very high temperatures generated; a *heat filter* in the beam path is highly desirable. A cooler beam of light is produced when the lamp is mounted in a *dichroic reflector* (see photograph on page 67); this is a sort of "selective valve" that allows much of the heating infra-red radiation to pass through whilst almost all of the shorter wavelengths of radiation are reflected back.

Light-Emitting Diodes (L.E.D.s)

These are genuinely miniature light sources and one can select from a range of colour options. Even within the "white" range there is variation in the quality of the light produced. Low voltage direct current is required and it is vital to ensure that the current passes through the L.E.D. in one direction only, as specified. Very little heat is generated. A great feature of L.E.D.s is that they can easily be organised into arrays, a notable one being the *ring-light*; if one of these can be positioned so that a ring of L.E.D.s surrounds the objective(s), a virtually shadow-less incident illumination is available (see image on p. 60).

Ignoring the lighting needs of specialised work, these are the four types of light source that are generally available:

Table of four lighting types

Tungsten	Operated at "mains voltage" and being phased out.
Energy Saver	Operated at "mains voltage" – replacing tungsten bulbs.
Tungsten-Halogen	Operated at low voltage (typically 6 V-12 V), usually from a "mains" supply through a transformer with voltage control. The possibility of using battery power can be important if portability is a requirement.
Light-Emitting Diode (L.E.D.)	Operated at low voltage and a direct current flowing in a specified direction is required. Suitable for portable battery power.

Light Sources and Their Spectra

Table of four lighting types, showing bulbs and their light spectra

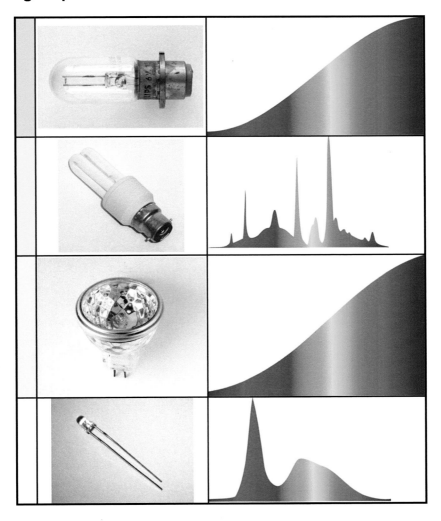

Watercolour Painting

Setting Up

The formulation of a strategy for examining a specimen with a stereomicroscope is usually not a difficult or time-consuming process. However, the adoption of a systematic approach will help to ensure that good results are achieved consistently.

You are now ready to go   **SOLO!**

The idea is to consider, in turn:

Specimen –   preparation and presentation.

Optics –   selection and insertion of optical elements
into the optical train.

Lighting –   positioning and switching on
of appropriate lighting units.

Observation –   adjustment for individual requirements
and good viewing.

There is a logical progression through a general planning process here but it is important to recognise that the issues involved are closely inter-related.

Let us now consider each of these matters, in turn.

SOLO Specimen

The following are the sort of questions to which it would be helpful to have answers:

What is the examination of this specimen to achieve?

Is my objective to:

Observe,
Measure,
Record images (by drawing, photography or digital capture),
Dissect or manipulate?

Do I wish to study:

Structure,
Textures,
Movement,
Dynamic processes such as crystal formation?

Setting Up

Examples of different samples:
e.g. Liquid samples (Pondwater), whole insect (Wasp),
mounted slide (Sundew), mineral specimen.

Setting Up

SOLO *Optics*

What optical elements will need to be placed in the
light path in order to achieve the effects required?

The combination of lenses to give an appropriate amount of
magnification of the object is the main consideration. In a
straightforward situation, with only (an) objective(s) and eyepieces in
use, an approximation of the total magnification for each eye is given
by:

Total Mag. = Mag. by Objective x Mag. by Eyepiece

The introduction of *supplementary lenses* and zoom mechanisms
would necessitate further adjustment of the result given by this
formula.

NOTE:

1. Resolution of fine detail, which is often of great importance
    when using a conventional compound microscope, is not
    generally an issue when working with stereomicroscopes; as
    we have seen, objectives in use are not designed to show
    fine detail and give high magnifications.

2. With Greenough type stereomicroscopes, the use of pairs of
    eyepieces that are not designed for use with a particular
    instrument can lead to difficulties. The paired objectives
    would not each have the same field of view after any
    necessary re-focusing to get their primary image in the
    optimum place for the eyepieces.

Setting Up

Examples of eyepieces and objectives:

From top left to right; a pair of eyepieces, black Greenough type objective, grey Greenough objective.
Bottom image; Common Main Objective (CMO).

Setting Up

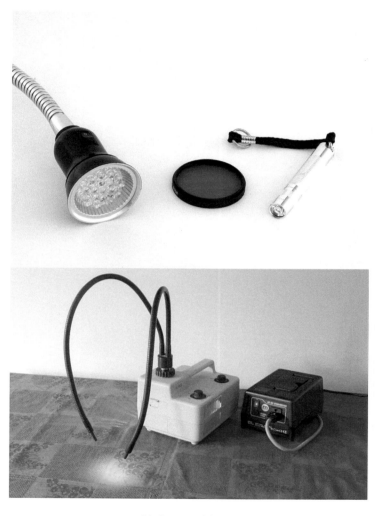

Lights and filter:
Top: Two different LED lamps and a green filter.
Bottom: Tungsten light with twin fibre optic light guides.

Setting Up

SOLO *Lighting*

What sort of lighting will be most helpful?

Try this.

Often, some compromises are required but, assuming that you have a full range of options available, decide which lighting arrangement to use in the following situations:

1. Examination of an opaque rock sample to view crystals in a cavity.

2. Observation of living "water fleas" (*Daphnia sp.*) in a small pool of pond water.

3. Measurement of the length of a bumblebee's tongue using a permanent slide preparation of a head with appendages.

4. Dissection of the stamens from a small flower to obtain pollen for mounting.

5. Photography of a stained section of plant stem.

Please turn to the next page to read some observations that will help you to evaluate your choices.

Setting Up

SOLO *Lighting continued*

The application of judgement and common sense is required. Often, there is no absolutely right or wrong answer.

1. Incident lighting is required. Any convenient light source could be employed; those which generate heat should not be put too close to the specimen. Oblique lighting would tend to create shadows and possibly enhance the 3-D effect.

2. Overheating of the animals must be avoided. Lighting through a *fibre-optic guide* would be ideal. L.E.D. sources would also be an acceptable choice. It would be worth experimenting with both incident and transmitted (including "*dark ground*") illumination.

3. Incident, or a blend of incident and transmitted light, would show up details to facilitate the alignment of the structure against a scale and the counting of divisions. Note; Such a scale, an *eye-piece graticule,* is engraved on a glass disc which is fitted in one of the eyepieces. Any convenient type of light source would be satisfactory.

4. Incident light is required. 'Balanced' lighting, i.e. from opposite sides or all-around, is helpful with a specimen that is being moved around. Any convenient source could be employed.

5. Is a conventional or digital camera to be used? The main consideration here is to select a light source that will give light of an appropriate quality. A stained specimen should have sufficient contrast but the use of filters to control quality of light might be helpful.

Setting Up

SOLO *Observation*

There are two things to concern us at this stage:

1. Setting-up and Adjusting

The manipulations for setting up a stereomicroscope are comparatively undemanding.

The "*working distance*" (the space between the front element of the objective lens and the object) is large.

One, "*coarse*" (cf. "*fine*") focussing knob is quite adequate to facilitate the viewing of a satisfactory image.

Fine adjustments to align optical components are not necessary.

(The one exception to this relates to alignment of the twin objectives in Greenough microscopes. With some instruments, objectives can be re-positioned slightly in their seatings by turning grub screws; however, the best advice is to ensure that the lenses, once adjusted properly, are not knocked out of alignment.)

There is less chance of damaging equipment or a specimen whilst setting up a stereomicroscope than whilst working with a compound instrument. However, the good standard procedures used when setting up any microscope should be employed. (see Check-list that follows on page 80).

2. Recording

Are the results of one's observations to be recorded and / or displayed? If so, a number of options are available.

Drawing.

Compilation of tables of measurements, graphs, etc.

Photography.

Video

The increased availability of computers and image-processing software has opened up all sorts of exciting possibilities for anyone wishing to capture their observations as high-quality images. There are several helpful web-sites which are worth exploring and photographic magazines contain useful advice.

Setting Up

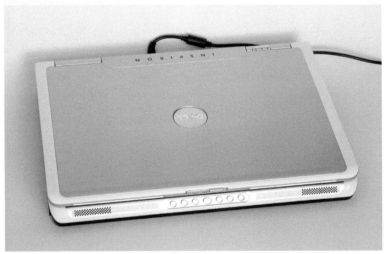

Different ways of recording your observations.
e.g. pencil and paper, camera or laptop.

Setting Up

A Check-list of Points to Remember When Setting Up

- Lift and carry instruments with one hand supporting the base.

- Work with a comfortable light intensity.

- Take care, when positioning specimens, to avoid damaging or contaminating the parts of the microscope, especially the optical components.

- Observe from the side whilst lowering the objective(s) to a position <u>below</u> that required. Then, whilst looking through the eyepieces, use the focussing knob to raise the head in order to obtain a clear image.

- Adjustment of the *inter-ocular distance* is achieved by carefully manipulating the eyepieces closer together or further apart until only one disc of illuminated field is visible. This is similar to the operation one performs when using field binoculars.

- Some instruments have one eyepiece (or both) that can be focussed independently to compensate for differences between the two eyes of an observer.

- If both eyepieces are adjustable, set the focussing ring on each to the mid-position of its range; next, use the focussing knob to obtain the best image possible with both eyes; then, view with one eye at a time and adjust the focussing ring on each eyepiece in turn until a sharp image is obtained. A final check with both eyes and use of the focussing knob, as necessary, possibly with small adjustments of lighting, should produce the optimum image.

- If only one eyepiece can be adjusted independently, start by viewing only through the non-adjustable eyepiece and use the focussing knob to obtain a sharp image; then, using only the other eye, adjust the focussing ring on the adjustable eyepiece until good focus is achieved. Final "tweaking" with the focussing knob and of the lighting, whilst viewing with both eyes, should then give a splendid image.

- (Note: Any measuring graticule present in one of the eyepieces should be superimposed on the primary image and would, therefore, be in sharp focus when proper adjustments have been made.)

- When using a stereomicroscope with a 'zoom' facility, one experiences tiredness if the eyes have to keep accommodating to images that are slightly out of focus. To avoid this, the microscope should be set up so that clear images are obtained across the range of magnifications.

Proceed as follows:

Start with the instrument zoomed in as much as possible. Set the eyepiece(s) to their mid position focus setting. Then, focus the microscope on a flat, contrasty specimen using the focussing knob. Next, take the zoom to its lowest position and get the clearest possible image by using the focussing rings on the eyepieces. Finally, go back to the highest zoom setting and, if necessary, make small adjustments to the focussing of the eyepiece(s) for an optimum image.

Setting Up

Care and Maintenance

- New instruments are usually supplied with rubber or plastic caps to fit on the eyepieces but these useful items can be acquired separately. The caps eliminate the possibility of spectacle lenses being scratched through contact with the top surfaces of the eyepieces.

- Some caps are adjustable, helping with the location of the eyes at the optimum level above the eyepieces. In this case, the top part of the cap can be raised or lowered. For unaided eyes, the top part is adjusted to the higher position while those wearing spectacles should snap or fold the top part of the cap down.

- A decent stereomicroscope is a valuable piece of equipment which will give many years of good service if it is respected. Care in using it and a little simple maintenance will pay dividends.

- When the instrument is not in use, a dust cover should be placed over it. In the absence of a dedicated cover, a clean polythene bag will serve very well.

- There is much advice in the literature about lens cleaning; some of this is conflicting! A safe approach is carefully to remove dust, hair, etc. with a soft brush and then wipe the accessible glass surfaces <u>gently</u> with a lens-cleaning tissue that has been <u>dampened</u> with a proprietary lens-cleaning fluid. One can then breathe heavily on the glass to moisten the surface and gently sweep a dry lens tissue over it again.

Setting Up

A Stephenson Microscope by James Swift and Sons.

Several different microscope systems, of which this is an example, were originally developed. Detailed information about these can be found in Carpenter & Dallinger, Hartley, the article by Nelson, and on various websites (see References and Further Reading).

Setting Up

Photography with the Stereomicroscope

Camera images taken through a microscope are called photomicrographs. Two sets of skills are required if one is to be successful in the creation of such images:

1. Setting up the microscope so that a high-quality image is provided. It has been said that this is the more challenging part of the work.

2. Image capture using photographic equipment.

Attention needs to be given to such things as the selection of appropriate lighting, exposure settings on the camera and any "zooming" of the camera to create the required field of view. In terms of the resolution of digital cameras, a 10 mega-pixel instrument is more than adequate for the work that we are considering, even if quite large prints are required; it is helpful to have a camera with a so-called "live-view" facility.

Over recent years, digital photography has virtually replaced the traditional techniques using film material and it is now easier for non-experts to achieve excellent results. It is assumed that the reader of this book will have developed the necessary understanding to set up a stereomicroscope properly; however, there is a vast amount of literature on "Microscopy" and a few, carefully-selected texts are suggested on page 93 for those seeking further guidance. Also included are some sources of guidance on the use of digital cameras.

There are alternative, recently-developed means of creating stereoscopic effects. Some so-called "stacking" programmes include a means of doing this. Also, there are lens attachments, even for mobile 'phone cameras, designed for this specific purpose.

The following notes are written on the assumption that images will be captured by cameras positioned above eyepieces in the relevant ports of a stereomicroscope; the camera effectively takes the place of the eye in direct viewing. It should be noted that there are other effective arrangements that require the eyepieces to be removed. We will not go into detail about these things here but will restrict ourselves to a consideration of those particular issues which need to be understood by the stereomicroscopist.

What Sort of Microscope Do You Have?

You may have a binocular or a trinocular head on your instrument. The latter has certain advantages for taking photographs.

With a trinocular head, there is usually a sliding or 'knob-turn' mechanism to divert the light from one (or, in rare cases, either) of the normal viewing light paths to the third port, often referred to as the photography port. There are also instruments in which the photography port shares an image with one of the normal viewing paths. The main advantages of having a third, photography port are:

- The camera is supported on a vertical tube and does not place the undesirable strain that is caused when a heavy camera is attached to an inclined eyepiece.

- It is easy to switch between the view through an eyepiece in use and the camera's view. These two images should be par-focal - both in focus at the same time; if this is not the case, some means of re-positioning the camera on its port will be necessary or re-focussing will be required when moving from direct viewing to ensure that the camera will capture a sharp image.

What Sort of Image Do You Want?

We have seen that it is necessary to take a pair of images if the observer of the end result is to see an image that has stereoscopic qualities; the images, one for each eye, are taken from slightly different viewpoints. A single shot, however taken, lacks the three-dimensional quality and will produce only the effect that is achieved by viewing a photomicrograph taken through a conventional compound microscope.

Do you want to create a stereo-pair?

No ....1. Single Shots (below)

Yes ....   2. Stereo Pairs (p. 90)

1. Single Shots

Your image can be captured through any of the available ports. The camera may be hand-held above the chosen eye-piece but there are numerous methods for supporting a camera so that it can be adjusted and held rigidly in the optimum position.

A Refinement for use with Greenhough-type Microscopes

On page 40, it was pointed out that, with this type of instrument, the optical axes of the two light paths are not perpendicular to the plane in which the object lies. This means that each image created will not be sharply in focus across the whole field of view. When viewing directly with both eyes, our eyes and brain are capable of "compensating" for this, so that the whole, 3-D image seen appears sharp. A camera cannot do this and a single shot, taken with either light path, will not produce an image that is sharply in focus across the entire field of view.

Photography with the Stereomicroscope

One way of overcoming this difficulty is to construct a "see-saw" mechanism.

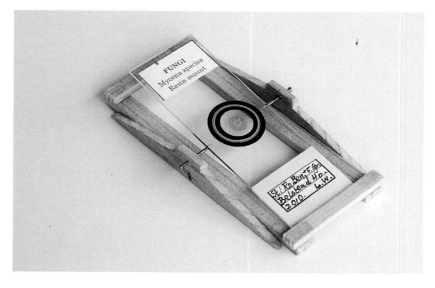

A see-saw mechanism for single shots with a Greenhough type stereomicroscope.

The specimen can be tilted so that either light path can be used effectively for taking a photomicrograph. The illustration above shows a mechanism that is suitable for specimens requiring incident ("top") and / or transmitted ("bottom") illumination. Photomicrographs 1 & 2 compare results obtained without and with the see-saw in use, respectively; look carefully and critically to see the difference, because your eyes and brain will try to give you what you expect or hope to see – the camera has made a faithful record!

Photography with the Stereomicroscope

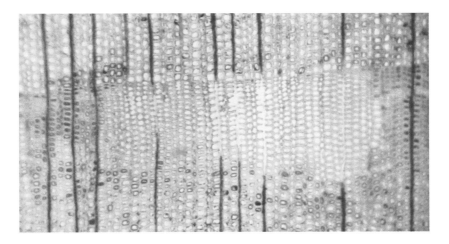

1. *Image taken with specimen flat on stage.*
Note blurring on left and right edges.

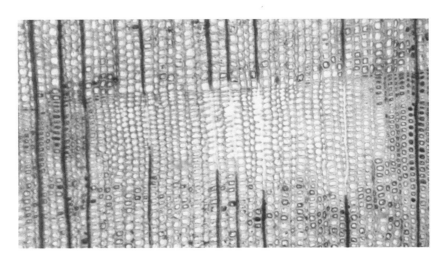

2. *Image taken with specimen on see-saw.*
Note image in focus across entire field of view.

Photography with the Stereomicroscope

2. Stereo pairs

Stationary Objects

As long as there is no accidental movement of the object, one has plenty of time to capture the two images required. These can be taken through the two eyepieces for normal viewing. Some microscopes allow the two separate light paths to be switched to the photography port in turn to be photographed individually there.

The trick here is to produce properly matched pictures. This means that the field of view needs to be the same for the two shots and the alignment needs to be the same in each case.

Moving Objects

Capturing a stereo-pair of images with a moving object is challenging and calls for some ingenuity. The two images required must be captured at the same instant.

Stereo-cameras are available. Such an instrument is essentially two identical cameras, side-by-side, built into a single case; a mechanism ensures that the images are captured by the two parts of the system simultaneously. If such a camera can be positioned so that its two lenses are correctly positioned above the viewing eyepieces of a pre-adjusted stereomicroscope, one could stand a good chance of achieving an acceptable result. A glance at the diagrams on pages 41 and 43 will help you to see why this arrangement is better suited to the use of C.M.O. stereomicroscopes; the axes of the light trains emerging from the eyepieces are in just the line required by the camera's twin lenses.

An alternative approach, which could be made to work well with any type of stereomicroscope, is to have a pair of matched cameras that are synchronised to 'fire' at the same instant. There is more scope here to ensure that each camera is positioned optimally over its respective eyepiece.

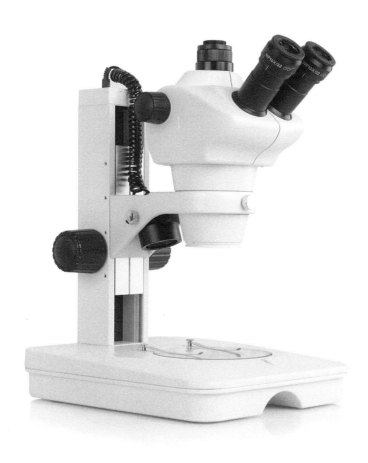

Aspen A1060 GTS6 Stereo Zoom Microscope.

From GT Vision, with permission.

Photography with the Stereomicroscope

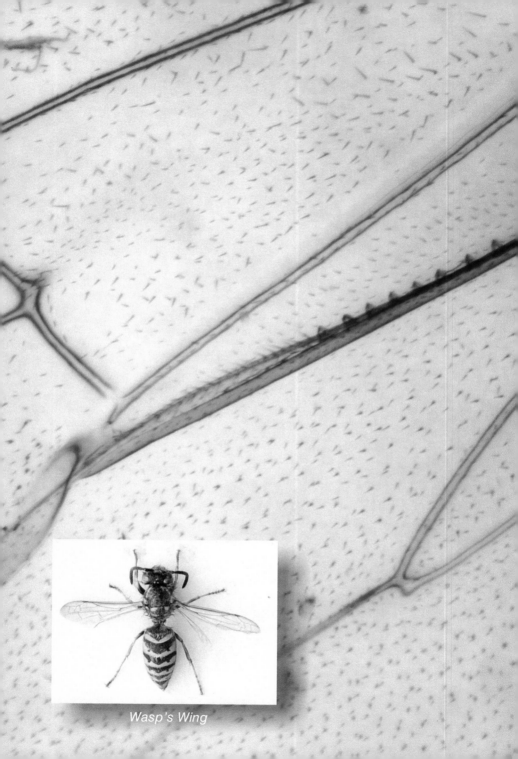

Wasp's Wing

References and Further Reading

General Microscopy

BRADBURY, S. & BRACEGIRDLE, B. 1998. Introduction to Light Microscopy. BIOS Scientific Publishers (R.M.S. Microscopy Handbook, 42). Oxford.

BRADBURY, S., EVENNETT, P.J., HASELMANN, H. & PILLER, H. 1989. Dictionary of Light Microscopy. O.U.P. (R.M.S. Microscopy Handbook, 15). Oxford.

CARPENTER, W.B. & DALLINGER, W.H. 1901 (8th. Edition). The Microscope, Vol. 1. (The Microscope and its Accessories). Churchill. London.

HARTLEY, W.G. 1983. The Microscope, A Basic Guide. Quekett Microscopical Club. London.

HARTLEY, W.G. 1993. The Light Microscope, Its Use & Development. Senecio. Oxford.

NELSON, E.M. 1914. Binocular Microscopes. Journal of Quekett Microscopical Club, Ser. 2, Vol.XII, No. 75.
Reproduced in: Microscopical Papers from the "Quekett". (Ed. B. Bracegirdle). Q.M.C. London, 1989.

OLDFIELD, R. 1994. Light Microscopy, An Illustrated Guide. Wolfe. London.

OLLIVER, C.W. 1947. The Intelligent Use of the Microscope. Chapman and Hall. London.

Photomicrography

MATSUMO, B. 2010. Practical Digital Photomicrography. Rocky Nook.

ROST, F. & OLDFIELD, R. 2000. Photography with a Microscope. C.U.P. Cambridge.

Making Stereo Pairs

FERWERDA, JAC. G. 2003. The World of 3D, A Practical Guide to Stereo Photography. 3-D Book Productions.

Websites

http://www.microscopyu.com/articles/stereomicroscopy/stereointro.html
Micscape, a free online microscopy magazine.
http://www.microscopy-uk.org.uk/

The Quekett Microscopical Club
http://www.quekett.org/

The Postal Microscopical Society
http://www.postal-microscopical-society.org.uk/

The Royal Microscopical Society
https://www.rms.org.uk/

The above organisations are a good starting point and these provide useful links to many others that are worth exploring.

References & Further Reading

Old £10 note

A Final Comment

Once the basics of stereomicroscope use have been understood and put into practice, there is much more excitement in store.

Examination of some specimens with polarised light can be very rewarding. The application of such techniques as "dark-ground" or Rheinberg illumination can give stunning results in appropriate circumstances.

The process of creating these, and other, interesting effects is much the same whether one is using a stereomicroscope or a conventional compound instrument. Guidance is to be found in text books on general microscopy, a selection of which is included in "References and Further Reading".

ENJOY YOUR MICROSCOPY!

Acknowledgements, 2018 Edition

I am grateful to Pam Hamer, Past President of the Quekett Microscopical Club, for her support and sound advice throughout this project and Phil Greaves, who co-ordinated the production of the original edition of this book. I thank them for their enthusiasm, advice and expertise.

Joan Bingley, Brian Bracegirdle, Alan Brinkworth, Maurice Moss and Don Thompson have also been excellent "critical friends". Barry Ellam, John Garrett and James Rider have made equipment available to be photographed for the illustrations and their comments have also been extremely useful. I am fortunate to have had the benefit of advice from a group of people with so much experience and understanding of the relevant issues. My patient and tolerant wife, Janet, has been a diligent proof reader and I thank her for the assistance and support she has given me.

Chris Thomas, for the humorous, challenging and often detailed discussions during the preparation of this edition, for helping with some of the illustrations and offering some of his photographs. We worked together as fellow microscopists! His company, Milton Contact Ltd., has also kindly taken over the publication of this book.

I thank Brunel Microscopes for kindly agreeing to allow our use of the photographs on pages 55 and 57 and GT Vision for permission to use their images on pages 60 and 91. Thank you, also, to the London Stereoscopic Company for permission to use a photograph of their Owl Stereoscope.

Whilst I am indebted to all those who have helped in any way with this project, responsibility for any deficiencies in the content of this material must rest with me.

L.W.